This **WALKER** book belongs to:

_ _ _ _ _ _ _ _ _ _

_ _ _ _ _ _ _ _ _ _

For my children, Alia, Zarena, Zaki and Jaan,
and for all the doctors and nurses and healthcare workers
who look after every bit of our brilliant bodies. **R.F.**

For my mother, who encouraged me to study medicine
but supported me in following my heart. **V.W.**

First published 2024 by Walker Books Ltd
87 Vauxhall Walk, London SE11 5HJ

This edition published 2025

2 4 6 8 10 9 7 5 3 1

Text © 2024 Roopa Farooki   Illustrations © 2024 Viola Wang

The right of Roopa Farooki and Viola Wang to be identified as author and illustrator respectively of this work
has been asserted in accordance with the Copyright, Designs and Patents Act 1988

This book has been typeset in Mali

Printed in China

All rights reserved.
No part of this book may be reproduced, transmitted or stored in an information retrieval system in any form or by any means, graphic,
electronic or mechanical, including photocopying, taping and recording, without prior written permission from the publisher.

British Library Cataloguing in Publication Data: a catalogue record for this book is available from the British Library

ISBN 978-1-5295-2358-4

www.walker.co.uk

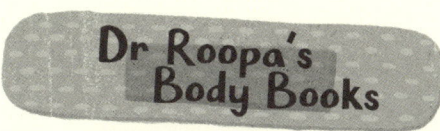

Dr Roopa's Body Books

# THE BRILLIANT BRAIN

**DR ROOPA FAROOKI**   ILLUSTRATED BY **VIOLA WANG**

WALKER BOOKS
AND SUBSIDIARIES
LONDON · BOSTON · SYDNEY · AUCKLAND

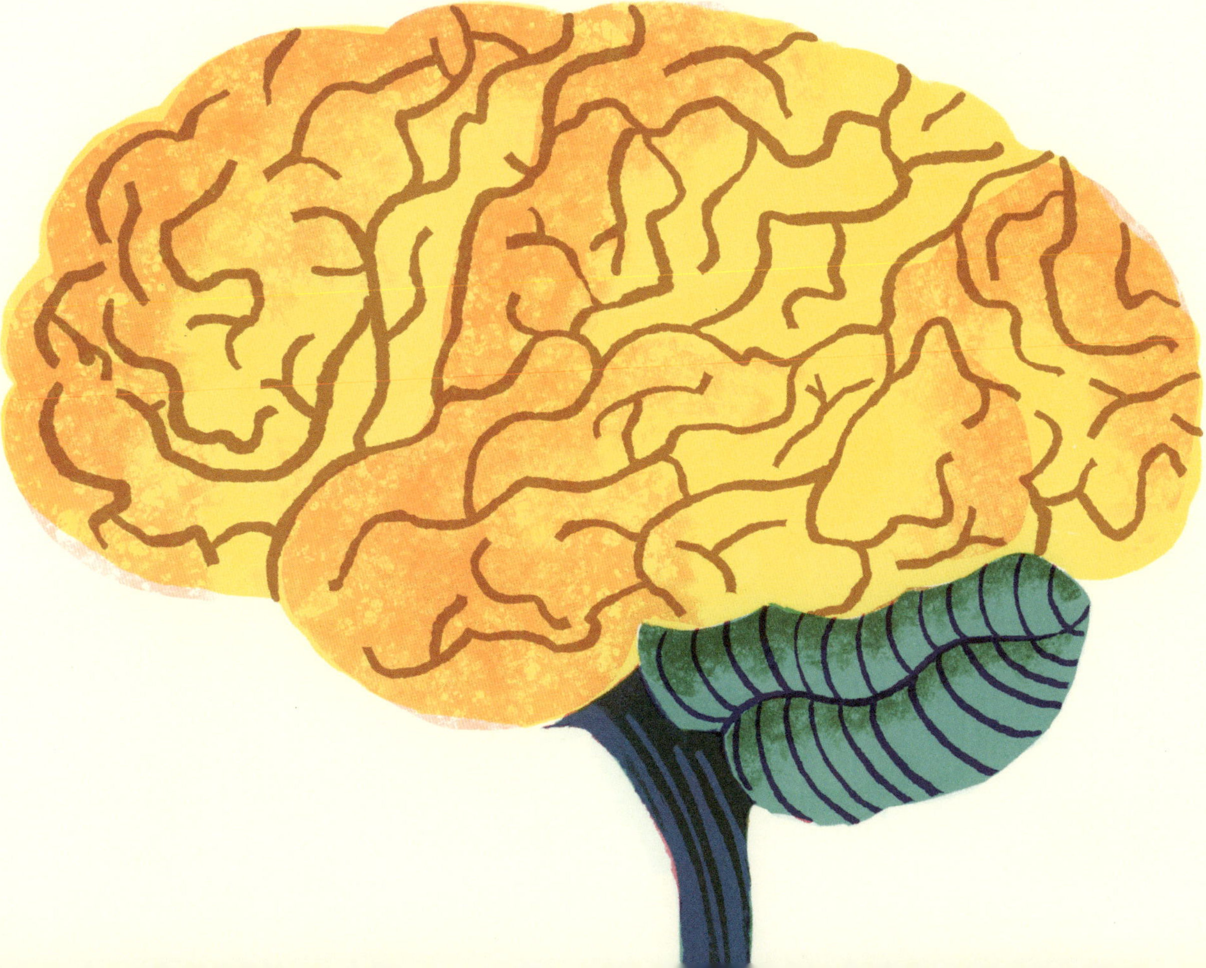

Every moment of every day, thoughts **POP** into our heads.

Sometimes, our thoughts are **pictures** ... sometimes, **words** ... sometimes, **feelings** ... sometimes, a bit of **everything**.

But where do we think our thoughts? The answer is: inside your own head. That's right, it's …

in your **BRAIN!**

Your brain is about the size of your two fists put together. It's dull and grey, soft and wet, with lots of squishy folds, and it doesn't really look very clever or impressive.

But if you think of your body as a kind of machine, your brain would be the control room — with MILLIONS of buttons and levers for doing all kinds of different jobs.

There's a different bit of the brain for EVERYTHING you think and feel and do. The largest part is called the "cerebrum" (*sur-ee-brum*). It's made of smaller parts called "lobes", which each send a particular kind of message around your body.

**FRONTAL LOBES** used for movement and decision-making

**TEMPORAL LOBES** used for hearing

# CEREBRUM

PARIETAL LOBES used for touching and feeling

OCCIPITAL LOBES used for sight (seeing with your eyes, or inside your mind)

So HOW do these messages get from your brain to the rest of you? The answer is ...

your **NERVES!**

BRAIN

NERVES

**SPINAL CORD**
a bundle of nerves that runs down your back

Your nerves spread out all over your body like branches, and carry instructions from your head to everywhere else.

The messages travel in both directions, to and from your brain. For example, this is what happens if you touch something pointy (like a cactus!).

The nerves in your skin send a message to your brain: *It's sharp! It hurts!*

Your brain then sends a message right back: *Ouch! Take out the spike!*

You often use several parts of your brain at the same time, even to do one simple thing.

Let's say you decide to go and say "Hi" to a friend. Your cerebrum tells your legs to get going...

But your legs also need to know HOW to move so you can keep your balance — and not wobble, or fall over!

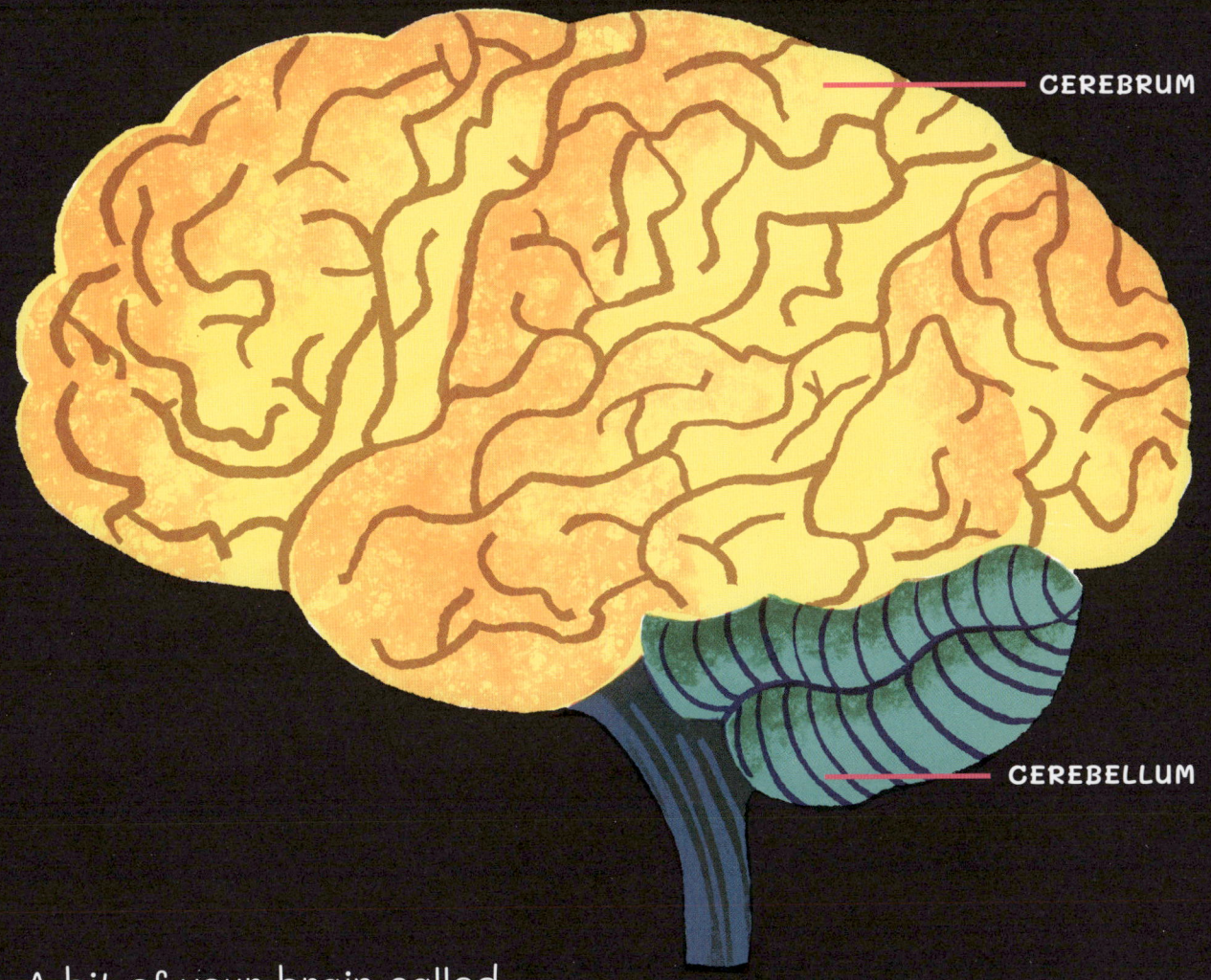

A bit of your brain called the "cerebellum" (*sair-ruh-bel-um*) helps different parts of your body to work together and move at the same time.

---

The cerebellum is also the bit of the brain which makes sure your lips and tongue are doing the right thing at the right moment when you actually say "Hi!"

There are some things your body does without you having to think about them — like your stomach breaking down what you had for lunch.

This is all thanks to your "brain stem". This bit of the brain also keeps blood pumping around your body — and even keeps you breathing!

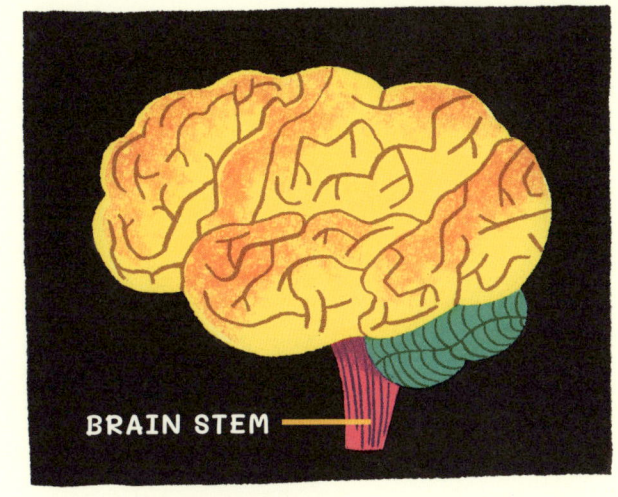

BRAIN STEM

If you hold your breath for a bit too long, the brain stem will take over and make you take a big gulp of air!

Your brain does so many other important jobs, too — like helping you to remember things.

Can you remember what you ate for lunch? Or where you left your favourite book? These are "short-term" memories, and they're stored in a part of the brain called the "hippocampus".

Let's have another think... Can you remember your own name? Or which foods you do and don't like? What silly questions! The answer is YES of course!

These important memories are stored all over your brain, to make sure you can't forget them easily.

Whenever you feel happy, sad or scared, and everything in between, your brain is hard at work, too.

If you're enjoying playing your favourite game, or laughing with your best friend, a part of your brain called the "limbic system" is busy sending you **happy** messages.

The limbic system also keeps you safe — when there's something dangerous nearby, it sends you **scared** messages and gives you a boost of energy to help you run away.

With some other emotions, like feeling **shy**, it's harder to tell where the messages begin — maybe they come from ALL AROUND your brain.

What's the single most special thing about your busy, bright, baffling brain?

The answer is ... you fill it, every day,
with everything you see,
feel and do.

So, it creates all the thrilling thoughts

and stores all the magical memories ...

that make **you YOU**.

## AUTHOR'S NOTE

My name is Doctor Roopa Farooki, and I've helped to look after every bit of people's brilliant bodies — from their brains to their big toes! I think one really important thing about our brains is that they help us make the right decisions: decisions that keep us healthy, happy and feeling great. Here are my tips for looking after your brain...

☆ **Try to eat a rainbow of different fruits and vegetables, and move your body when you can.**

✭ Drink lots of water — your brain is three-quarters water, so you need to drink plenty to help keep it working well.

✭ You can give your brain lots of fun challenges to keep it active — like playing games with your friends, making art, solving puzzles and building things.

✭ Watch out for anything that might hurt your head! Always wear a helmet if you're riding a bike or doing some sort of sports activity where you could fall.

✭ Get lots of sleep — your brain finds it hard to focus if you're tired.

Also in this series:

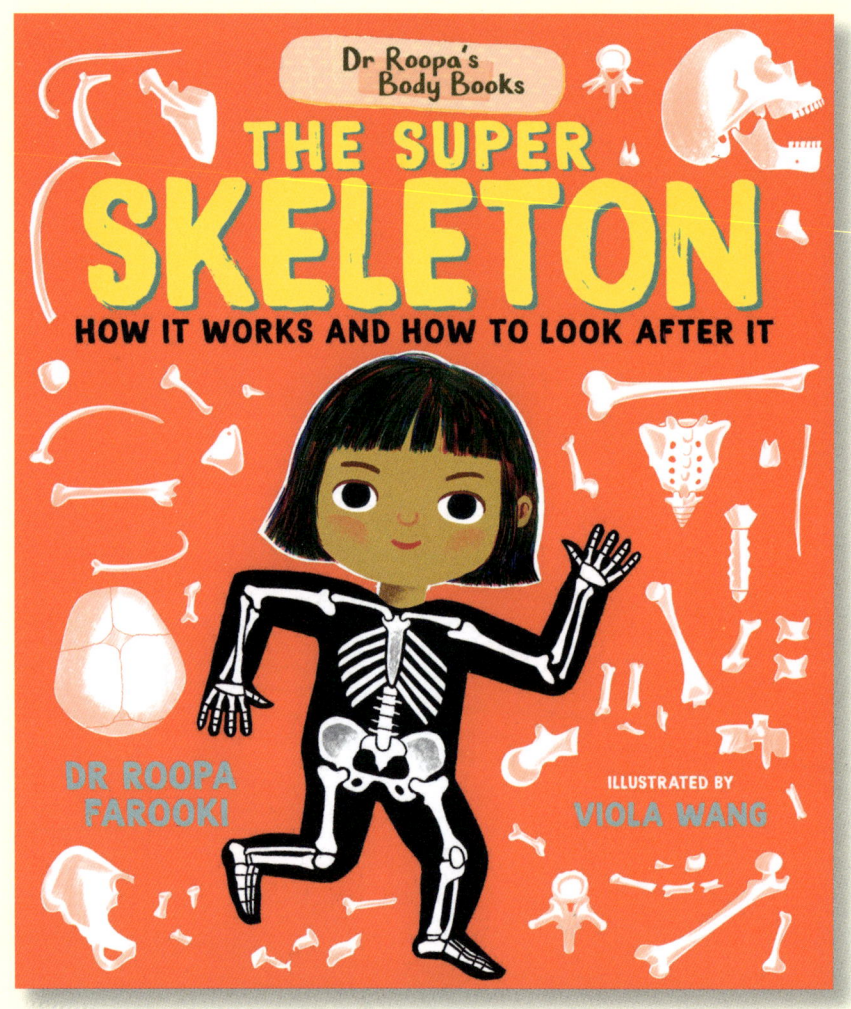

ISBN 978-1-5295-0452-1

Available at all good booksellers

www.walker.co.uk